I0715301

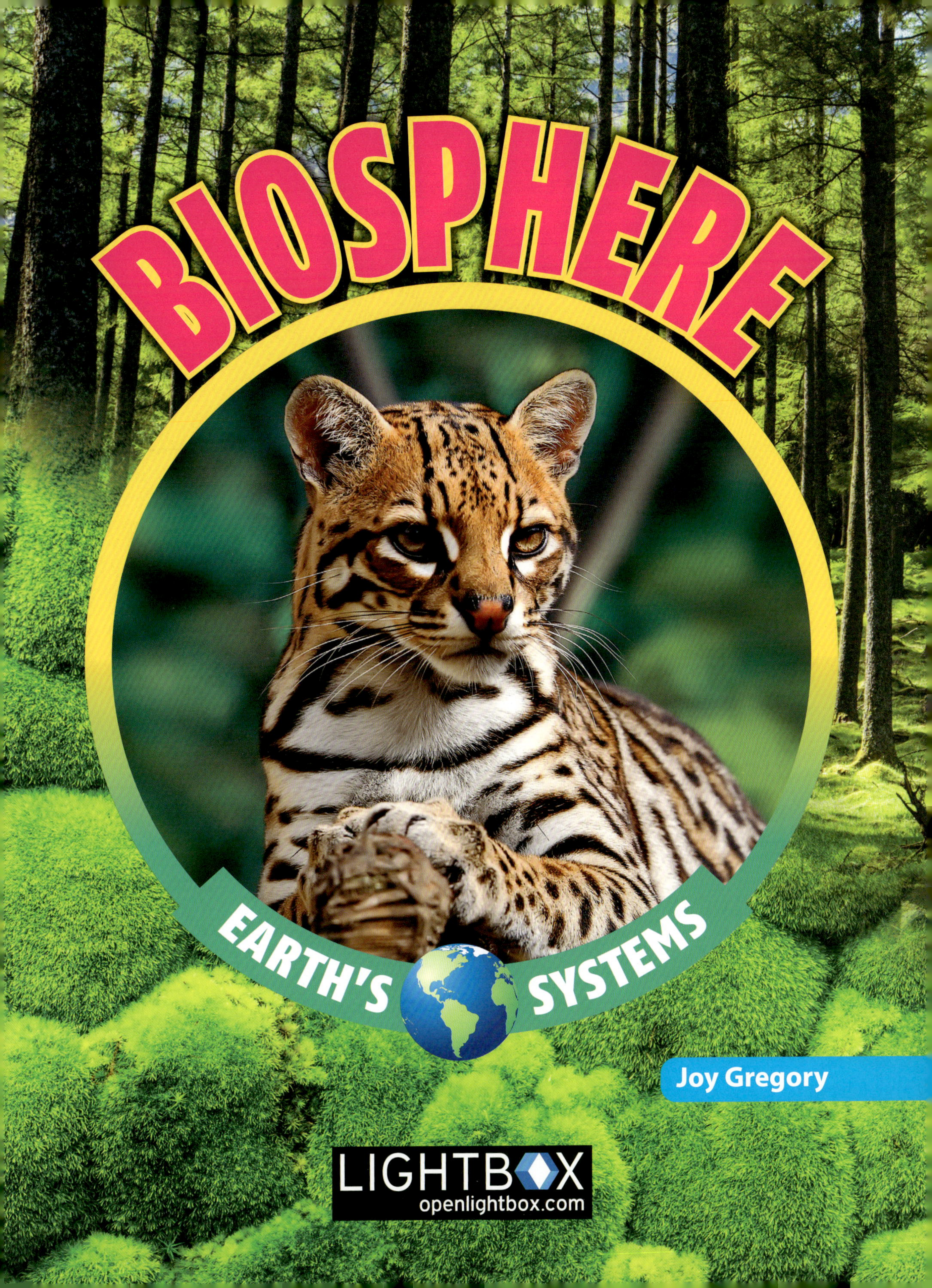

BIOSPHERE
EARTH'S SYSTEMS
Joy Gregory
LIGHTB X
openlightbox.com

LIGHTBOX

Go to
www.openlightbox.com
and enter this book's
unique code.

ACCESS CODE

LBXP4926

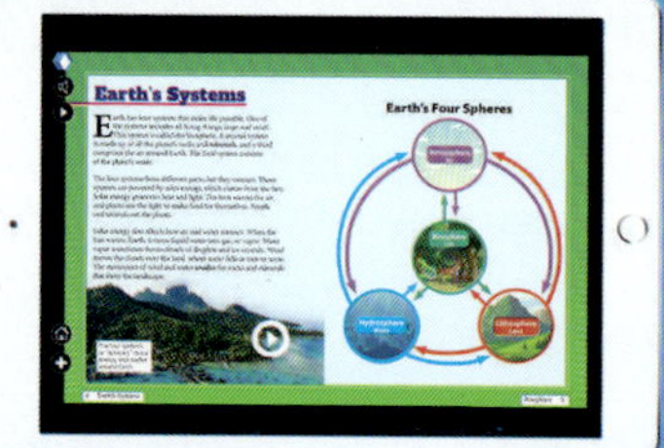

Lightbox is an all-inclusive digital solution for the teaching and learning of curriculum topics in an original, groundbreaking way. Lightbox is based on National Curriculum Standards.

LIGHTBOX SUPPLEMENTARY RESOURCES

SHARE
Share titles within your Learning Management System (LMS) or Library Circulation System

CURRICULUM
Find national and state curriculum correlations

CITATION
Create bibliographical references following the Chicago Manual of Style

STANDARD FEATURES OF LIGHTBOX

AUDIO High-quality narration using text-to-speech system

ACTIVITIES Printable PDFs that can be emailed and graded

SLIDESHOWS Pictorial overviews of key concepts

VIDEOS Embedded high-definition video clips

WEBLINKS Curated links to external, child-safe resources

TRANSPARENCIES Step-by-step layering of maps, diagrams, charts, and timelines

INTERACTIVE MAPS Interactive maps and aerial satellite imagery

QUIZZES Ten multiple-choice questions that are automatically graded and emailed for teacher assessment

KEY WORDS Matching key concepts to their definitions

BIOSPHERE

CONTENTS

Earth's Systems

arth has four systems that make life possible. One of the systems includes all living things, large and small. This system is called the biosphere. A second system is made up of all the planet's rocks and **minerals**, and a third comprises the air around Earth. The final system consists of the planet's water.

The four systems have different parts, but they interact. These systems are powered by solar energy, which comes from the Sun. Solar energy generates heat and light. The heat warms the air, and plants use the light to make food for themselves. People and animals eat the plants.

Solar energy also affects how air and water interact. When the Sun warms Earth, it turns liquid water into gas, or vapor. Water vapor sometimes forms clouds of droplets and ice crystals. Wind moves the clouds over the land, where water falls as rain or snow. The movement of wind and water **erodes** the rocks and minerals that form the landscape.

The four systems, or "spheres," move energy and matter around Earth.

Earth's Four Spheres

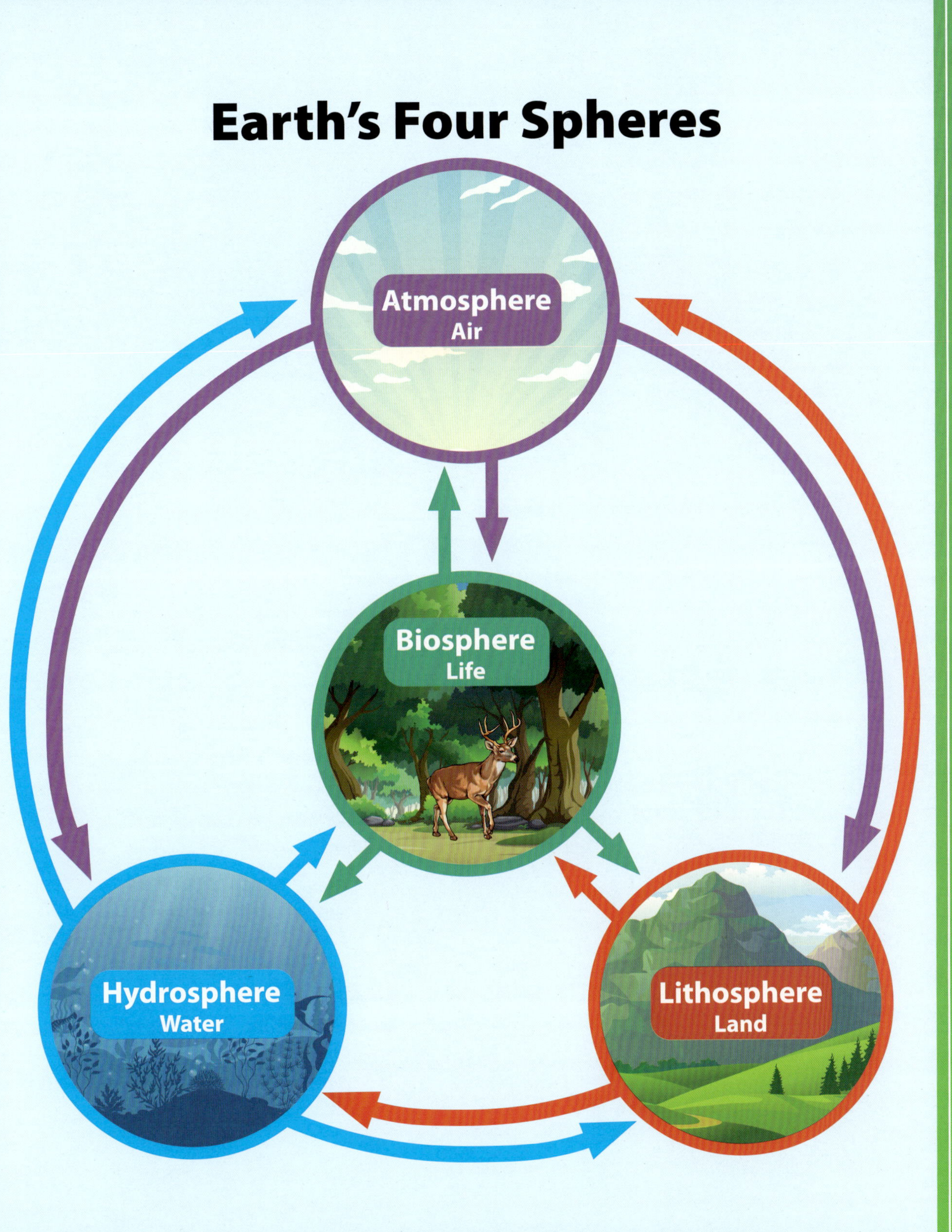

Animals and plants in particular regions of the biosphere all interact.

The Biosphere

The biosphere consists of all living things on land, in the air, and in the water. This sphere includes humans, which are the most complicated organisms on Earth. Humans have about 30 trillion cells. The simplest organisms have only one cell. These **microorganisms** often live inside other living things.

The biosphere overlaps with Earth's other three systems. It stretches from far up in the atmosphere to the soil beneath Earth's **crust**, where the deepest tree roots grow. Some single-celled organisms live in the Mariana Trench, which is the deepest part of the ocean.

Many organisms can only live in certain parts of the biosphere. These plants and animals have **adapted** to the conditions in specific **habitats**. Others require a particular kind of food. Pandas, for example, must live where they can find bamboo, their main food. Some organisms need access to different habitats. Salmon live in salt water but lay their eggs in fresh water. A few animals, such as frogs and otters, live both on land and in water.

Changes to one part of the biosphere can cause problems in other parts of the system. **Climate change** causes extreme weather events, including tornadoes, hurricanes, and droughts. Some of these events cause floods, along with wind and water erosion. This damages habitats and changes the biosphere.

Six Kingdoms of Life

The biosphere is divided into six large kingdoms of living things. Animals and plants form the largest kingdoms. Fungi include life forms such as mushrooms and mold. The other three kingdoms are made up of simple, single-celled life forms.

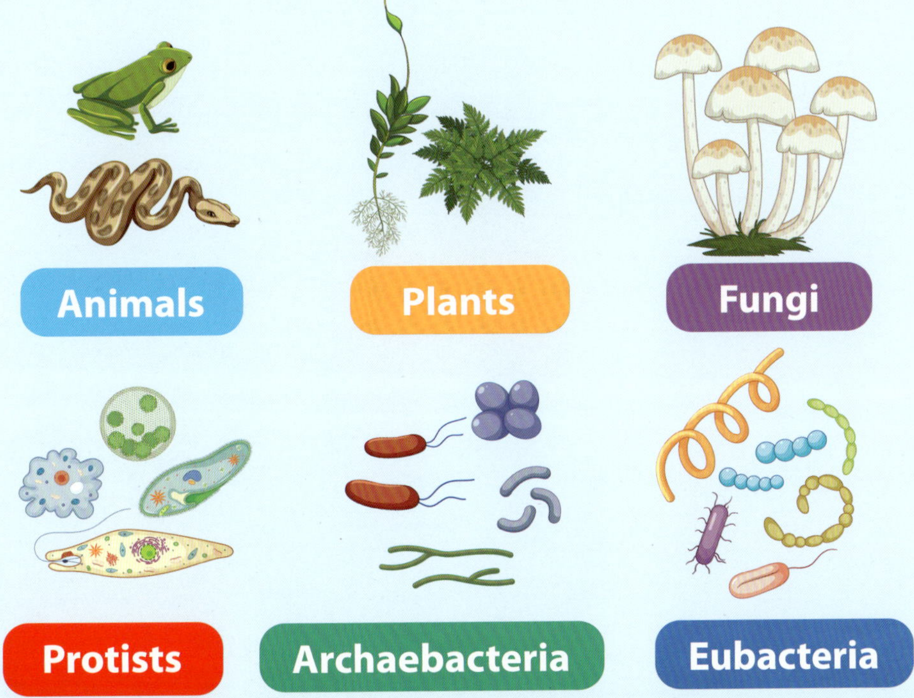

Levels of Organization

Every organism in the biosphere eats, lives, and dies within the biosphere. This puts the biosphere in a state of constant change. The biosphere is organized in different levels that relate to one another.

Every organism in the biosphere exists as a single individual. Individuals can be grouped into populations, or groups of the same kind of organisms, such as wolves or vultures. Populations of organisms live in communities. Ants, for example, form a colony. Humans form a city.

Habitats are like neighborhoods that provide homes for specific individuals, populations, and communities of organisms. Different habitats form biomes, which have plants and animals that share common traits, or characteristics. These traits allow plants and animals to survive in particular climates. Every biome includes different habitats. Together, organisms and their habitats form the biosphere.

The Food Chain

Organisms are connected by what they eat. This is called a food chain. Each food chain has producers and consumers. Plants are producers. They use energy from the Sun to make food. Animals are consumers. They cannot make their own food. Herbivores are animals that eat plants, while carnivores eat other animals.

2. Population
3. Community
1. Organism
6. Biosphere
4. Habitat
5. Biome
Sun
Producer
Herbivore
Carnivore

Around the World

The biosphere is easily damaged. It can be impacted by changes to the air, water, or land. In 1976, the United Nations (UN) set up its first biosphere reserve. These reserves are protected areas. Today, the UN has nearly 700 biosphere reserves in 129 different countries.

1 Haleakala National Park, North America

Haleakala National Park is located on the island of Maui, Hawaii, in the Pacific Ocean. It became a biosphere reserve in 1980. The park has a **dormant** volcano, forests, and beaches. Haleakala has more endangered species than any other U.S. national park. It is one of 28 UN biosphere reserves in the United States.

2 Julian Alps Biosphere Reserve, Europe

The Julian Alps Biosphere Reserve is in the mountains between Italy and Slovenia. It is one of 21 biosphere reserves that are located in more than one country. The reserve includes Mount Triglav, which is the highest mountain in Slovenia. Rare species of animals and flowers live in this protected area.

3 Yangambi Biosphere Reserve, Africa

Yangambi was one of the first biosphere reserves in the world. It is located in the Democratic Republic of the Congo, in Central Africa. Yangambi has been protected since 1976. The reserve contains about 32,000 different species of trees. Several endangered animals live there, including elephants and river hogs.

Grizzly bears eat salmon migrating along rivers.

Living with the Biosphere

Life in the biosphere is a constant competition. Organisms compete for the resources that keep them alive, including air, water, and land. Plants compete with other organisms for water and minerals. Some animals eat other animals. Microbes feed on plants, animals, and other microbes.

Many organisms have adapted to changes in resources. These changes include the ones caused by climate. Birds, for example, may **migrate** to other biomes when the seasons change. Other animals, such as grizzly bears, **hibernate** when the weather turns cold. Organisms that do not adapt may die because they do not have enough food, air, or water. Some organisms go **extinct**.

One of the worst sources of air pollution is smoke from factories.

Life in the biosphere is impacted by pollution. Pollution can be caused naturally, but most is created artificially by people. In the air, pollution creates smog. This toxic gas replaces oxygen in the air. Smog also blocks sunlight before it reaches plant leaves, and it traps heat in the atmosphere. The trapped heat changes living conditions in the biosphere. It causes climate change when it warms Earth's air and water.

Biosphere 2

Biosphere 2 is Earth's largest artificial biosphere. Located in Arizona, Biosphere 2 makes its own food and oxygen. The site opened in 1991. It was used to study how people could live in an artificial environment. Biosphere 2 is still used for research, but now visitors can also tour the site.

Human Activity

All human activity occurs in the biosphere. Many of the other organisms in the biosphere are also essential to human activity. They provide the raw materials people need to stay alive. Much of the global economy is based on how people use other living things to meet the need for food, clothing, and shelter.

Cereal crops, such as wheat and corn, are widely eaten in many countries.

Many plants and animals in the biosphere provide food for people. In the past, people gathered wild plants for food, or hunted wild animals. Today, most of the food people eat is grown on commercial farms and ranches. In some parts of the world, people use horses and oxen for transportation, and to plow fields.

Some plants, such as cotton and flax, are grown to make clothing. Clothing can also be made from sheep's wool and animal hides. Over time, people learned that certain plants could make them feel better if they were sick. They started to make tea from willow bark to soothe headaches, and tea from ginger root to help a sore stomach. These discoveries led to the development of the pharmaceutical industry. This business uses components of living organisms to make medicines, including **antibiotics**.

Cotton fibers are woven into textiles to make clothing.

The biosphere provides the materials people need to survive and be comfortable in Earth's climate. People cut down trees to build houses. Other trees are planted to shield homes from the Sun's rays. People use products developed in the biosphere to light and heat homes and businesses, too. Coal, oil, and **natural gas** come from organisms that used to live in the biosphere.

Timber is an inexpensive building material that is widely used around the world.

Human Impact

Fossil fuels power most cars and trucks.

Humans impact every living organism in the biosphere. Most of the negative impacts are the result of activities that create waste. This waste pollutes the air, land, and water. Pollution makes habitats toxic to living things.

Air pollution can be caused by dust, volcanoes, and wildfires. However, it is mainly generated by burning **fossil fuels** to make the electricity that lights, heats, and cools homes and offices. Air pollution collects in the atmosphere. It irritates people's lungs, causing allergic reactions and making people cough. Polluted air also holds less oxygen, making it harder to breathe.

Rain carries pollution from the air to the ground. The rain washes waste into the water, and the waste then pollutes the soil. People also pollute the land and water with garbage and sewage, which contaminates wildlife habitats. Pollution can make water unsafe for people to drink. It also ruins the soil and water that people use to grow food. People get sick when they eat food crops that are contaminated by sewage and chemicals.

Seabirds caught in oil spills die if their feathers are not cleaned quickly.

Reduce, Reuse, Recycle

In the United States, people create about 292 million tons (265 million metric tons) of garbage each year. That equals 5 pounds (2.3 kilograms) of garbage per person every day. Much of this garbage goes to landfills or is burned. The U.S. Environmental Protection Association urges people to create less waste. To do this, people can buy less, reuse more of what they own, or recycle the items they buy. The three Rs—reduce, reuse, and recycle—protect the biosphere by cutting waste.

Engineers have developed solar panels to turn energy from sunlight into electricity.

Science and Progress

Science is helping the biosphere recover from the damage caused by pollution. One way it is doing this is by using **renewable energy**. This technology cuts **greenhouse gas** pollution. Fewer greenhouse gases mean less toxic waste in the air. This helps keep the water and land cleaner, which is good for human health. Less pollution also reduces climate change.

Aquatic environments in the biosphere need attention, too. People can avoid putting raw sewage in rivers and oceans. This helps biomes grow healthier communities of plants and animals. When aquatic biomes are less healthy, they absorb more carbon dioxide, which heats up the water. The water evaporates and warms the air above it. This warming can cause high rainfall that creates problems by disrupting normal weather patterns.

The problem of plastic is another one that science is tackling. Plastic garbage can trap animals when it gets wrapped around their wings and legs. Land and aquatic animals that eat plastic waste can be poisoned by chemicals in the plastic. Scientists are coming up with new forms of packaging that use fish waste or other materials instead of plastic. These materials are biodegradable, which means they break down naturally after use.

People are also finding new ways to deal with plastic trash. Communities organize clean-up days for beaches and wetland areas. Other groups collect floating islands of plastic. The plastic is recycled and reused.

Trash that does not degrade naturally has to be cleared from beaches by hand.

The Biosphere through Time

The first organisms on Earth lived in water. Over millions of years, they released oxygen into the water and the air. As the atmosphere changed, more complex organisms developed. Plants such as mosses, along with bugs and amphibians, came first. Mammals, birds, and flowers followed.

1973

The U.S. Congress passes the Endangered Species Act. This law protects plants and animals and preserves important habitats.

2019

Hawaii's nene goose moves off the endangered species list. Since 1967, its population has risen from 30 to more than 2,800 birds.

2020

Panna National Park in India is declared a UN biosphere reserve. The park includes critical tiger habitat.

The Biosphere and the Atmosphere

The atmosphere is made up of the air around Earth. This air, which is mostly nitrogen and oxygen, is arranged in layers around the planet. Gases grow less dense farther above the ground. Air is thickest in the troposphere, the layer closest to Earth's surface. It grows thinner farther out from the planet. Only the troposphere can support life, and it is where the biosphere exists.

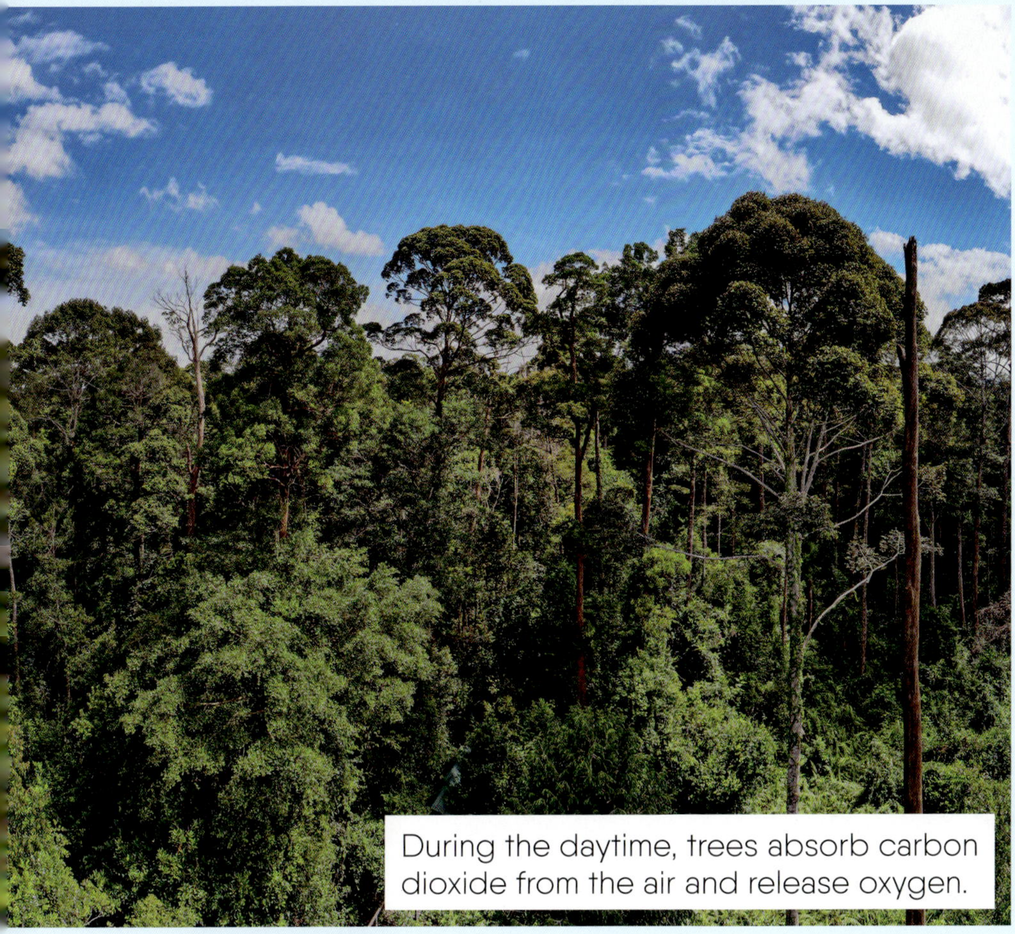

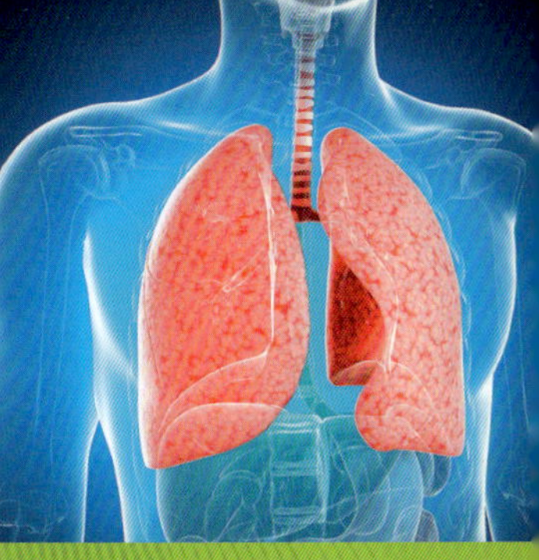

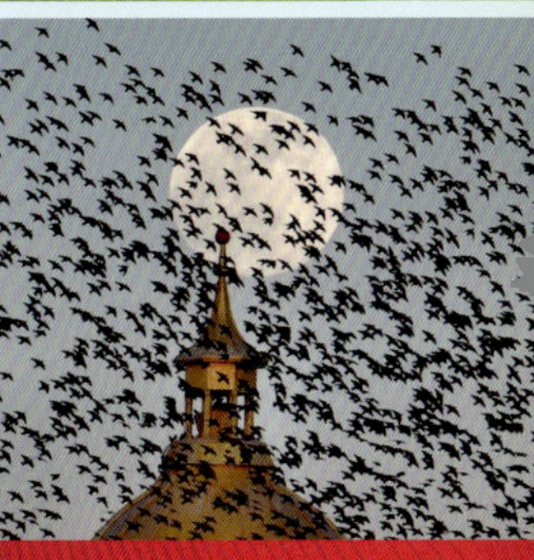

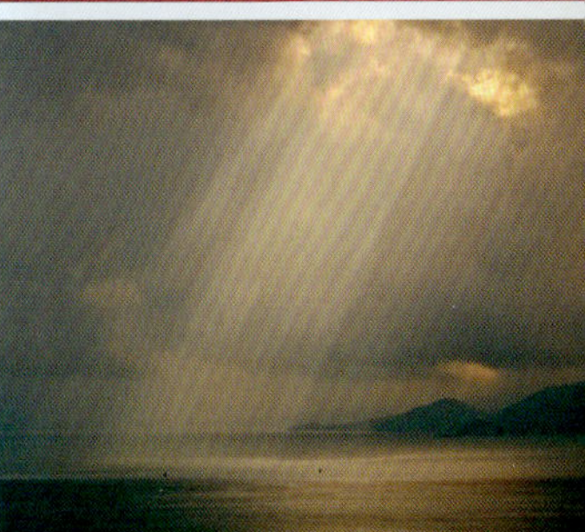

During the daytime, trees absorb carbon dioxide from the air and release oxygen.

Life in the biosphere depends on the atmosphere, which provides oxygen for people and animals, as well as a habitat for birds. The atmosphere absorbs or reflects the Sun's rays. This keeps Earth from becoming too hot or too cold, and prevents direct sunlight from burning plants and animals.

The atmosphere also protects the biosphere from damage caused by space objects. Rocks from space sometimes hit Earth. In the past, these collisions have killed many plants and animals. Most space rocks burn up in the atmosphere, however, due to **friction** with particles of gas.

Oxygen, which makes up **21 percent of the atmosphere**, is used by most **living organisms.**

Most **birds** fly in the lowest **1,640 feet** (500 meters) of **Earth's atmosphere**.

The **atmosphere absorbs** or **reflects** about **90 percent** of the **Sun's** most harmful **rays**.

The Biosphere and the Hydrosphere

Many of Earth's animals are aquatic, which means they live in the hydrosphere.

The hydrosphere includes all the water on Earth. Water can be found in solid, liquid, or gaseous forms. Solid water is called ice, and water in its gaseous form is called water vapor. Most of Earth's water is liquid, which is critical to the biosphere. Organisms need water to survive. Many plants and animals live in the water, which covers about 71 percent of Earth's surface. Water vapor occurs in the atmosphere.

Almost all the liquid water in the hydrosphere is salt water. Salt water is found in some lakes, but most of it is in the ocean, which is called the marine biome. This biome is different from the freshwater biome, which includes rivers, creeks, ponds, and lakes. People, together with a wide range of plants and animals, need fresh water to survive. A few types of fish, including salmon and trout, can live in both salt water and fresh water.

Organisms that live in water have adapted to their biomes. Both saltwater and freshwater fish have fins to help them move through water. Some marine mammals, such as seals and whales, have blubber, a type of fat. Blubber helps these animals stay warm in cold water. Fish have gills to breathe in the water.

Only **1 percent** of all of the water on Earth is **fresh water**.

The **ocean** holds about **96.5 percent** of the **water** on Earth.

Most kinds of **ocean life** live within **650 feet** (200 m) of the water's surface.

Fish gills absorb oxygen from water and release carbon dioxide.

The Biosphere and the Lithosphere

All of the rocks and minerals on Earth are part of the lithosphere, which provides a home for many living things in the biosphere. Marine and freshwater biomes exist in basins made of rocks and minerals. Land habitats have mountains, sand, and soil.

Farmers use soil that is rich in nutrients to grow food crops.

The lithosphere began to take shape millions of years ago. Below Earth's crust, **tectonic plates** of rock shifted, causing earthquakes and volcanoes. These events created mountains and valleys where plants and animals would eventually live. Water settled in low areas, and **glaciers** pushed rocks over the landscape. As time passed, rocks and minerals eroded into small particles that formed soil in which plants could grow. Different kinds of plants adapted to different soil and weather conditions in the biosphere. These plants created habitats for other plants and animals.

Organisms in the biosphere constantly change the lithosphere. Plant roots break apart soil and stone. Animals ranging from insects to bears dig holes for food and shelter. People cover parts of the land with roads and cities. Farmers dig trenches so water can irrigate crops. Engineers build railroad tracks through mountains. People also extract materials from the lithosphere, including minerals, such as limestone and clay, and diamonds and gold that are used in jewelry.

Termites live in tall mounds that they build by piling up grains of soil.

Shifting **tectonic plates** cause **Mount Everest** to grow about **16 inches** (40 centimeters) a century.

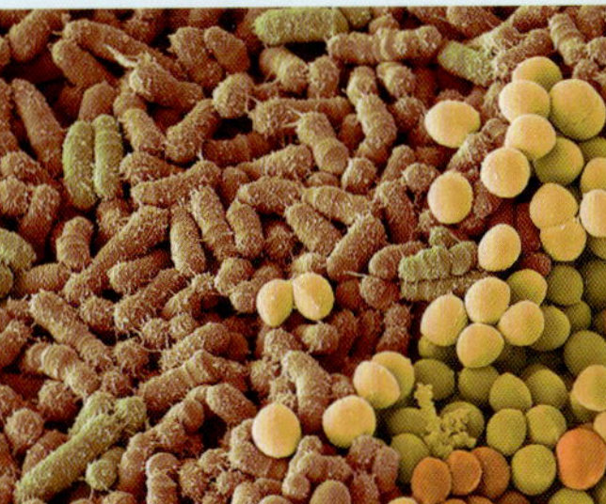

Microbes were found in **volcanic rock** buried by **870 feet** (265 m) of sediment.

At **33,500 feet** (10,210 m), **Mauna Kea** in Hawaii is the **world's tallest mountain** from base to peak.

Build a Model Biosphere

This activity lets you study how the biosphere interacts with the hydrosphere. It tests how plants use water and energy from the Sun in order to grow. Collect the materials you need and follow the instructions to begin. This experiment takes place over a few weeks.

All living plants depend on water in order to grow.

BEFORE YOU START

Experiments help us understand the world. They teach people how things do and do not work. Scientists use experiments to learn how to make things work faster, or better.

What You Need

2-cup glass jars

4 bean seeds

Ruler

Paper and pen

2 paper
towel sheets

Sunny counter

Dark cupboard

Water

What You Do

1. Fold each paper towel sheet in half and roll it into a cylinder.

2. Place a paper cylinder in each jar.

3. Adjust the paper so it touches the sides of the jar.

4. Put enough water in the jars to wet the paper towel.

5. When the paper towel is damp, drain extra water from both jars.

6. In each jar, place two bean seeds between the glass and the damp paper towel.

7. Place one jar in a sunny spot on the counter. Put the other jar into a dark cupboard.

8. Check the jars every day, and add enough water to keep the paper towels damp.

9. Measure how the roots and shoots grow for a few weeks.

Questions

1. Did the seeds germinate at the same time? Why?

2. Which jar developed green leaves first?

3. What role did sunlight play in plant growth? Why?

Biosphere Quiz

1 Who first used the term "biosphere"?

2 How many UN biosphere reserves are located in the United States?

3 Which U.S. national park is home to more endangered species than any other national park in the country?

4 Which gas do fish gills absorb from water?

5 About how many years ago did life on Earth begin?

6 How many countries have UN biosphere reserves?

7 Approximately how many species of trees are found at the Yangambi Biosphere Reserve in the Democratic Republic of the Congo?

8 Which gas do trees absorb from the air during the daytime?

9 How much of all the water on Earth is fresh water?

10 In what layer of the atmosphere does the biosphere exist?

ANSWER KEY

1. Eduard Suess **2.** 28 **3.** Haleakala National Park **4.** Oxygen **5.** 3.5 billion **6.** 129 **7.** 32,000 **8.** Carbon dioxide **9.** 1 percent **10.** Troposphere

Key Words

adapted: changed to live more easily in particular conditions

antibiotics: medicines that treat infection

climate change: a gradual change to Earth's weather patterns over time, often caused by human activity

crust: the outer layer of Earth's surface

dormant: asleep or not active

erodes: breaks down or wears away natural features through wind or water

extinct: no longer existing

fossil fuels: energy sources made from coal, wood, oil, or natural gas

friction: the force created when two or more things rub against each other

glaciers: huge masses of ice that move slowly over land

greenhouse gas: a gas, often created by burning fossil fuels, that traps heat in the atmosphere

habitats: the places where organisms live

hibernate: to become inactive, usually during the winter

microorganisms: single-celled organisms such as bacteria and germs

migrate: to move from one region to another as the seasons change

minerals: solid substances found in the ground that are often useful, such as salt or potassium

natural gas: a gas found underground that can be burned as a fuel

renewable energy: energy created from resources, such as wind, water, and the Sun, that do not run out

tectonic plates: large plates of rocks deep under the ground that are always shifting

Index

LIGHTBOX

➕ SUPPLEMENTARY RESOURCES

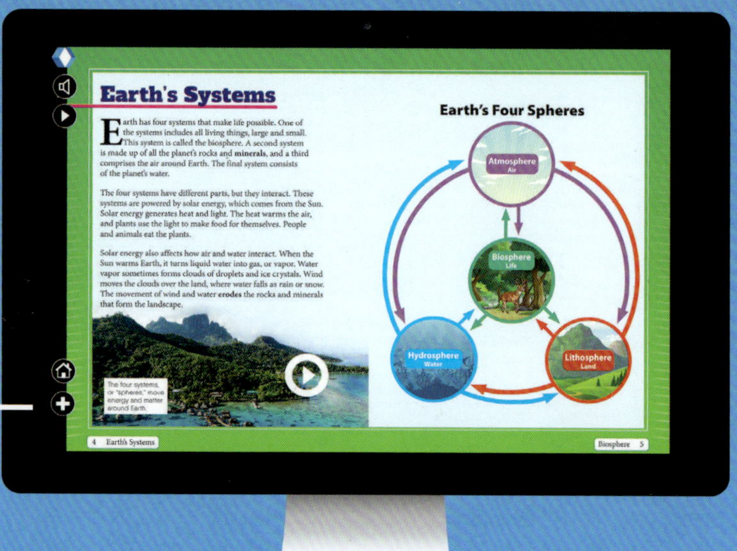

Click on the plus icon ➕ found in the bottom left corner of each spread to open additional teacher resources.

- Download and print the book's quizzes and activities
- Access curriculum correlations
- Explore additional web applications that enhance the Lightbox experience

LIGHTBOX DIGITAL TITLES
Packed full of integrated media

VIDEOS

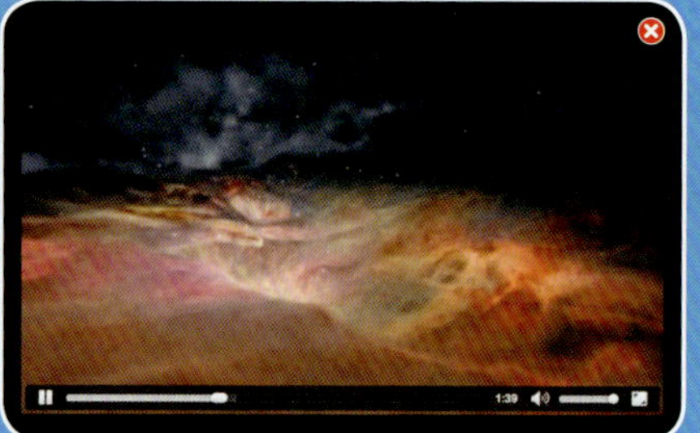

INTERACTIVE MAPS

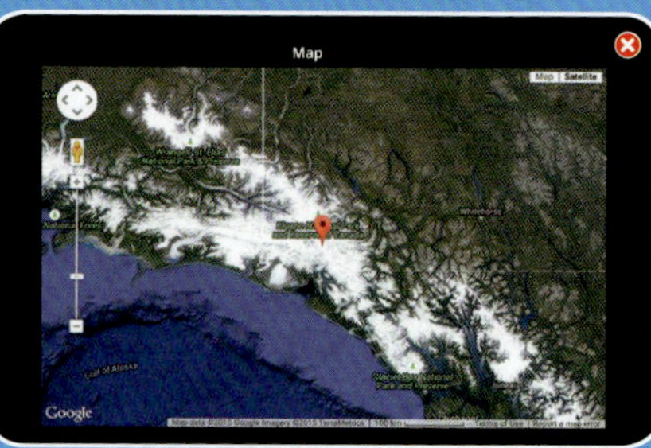

WEBLINKS

SLIDESHOWS

QUIZZES

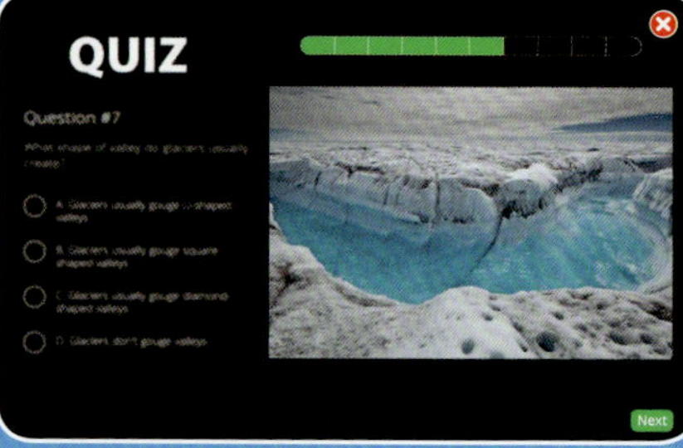

OPTIMIZED FOR

✓ **TABLETS**

✓ **WHITEBOARDS**

✓ **COMPUTERS**

✓ **AND MUCH MORE!**

Published by Smartbook Media Inc.
276 5th Avenue, Suite 704 #917
New York, NY 10001
Website: www.openlightbox.com

Library of Congress Cataloging-in-Publication Data

Names: Gregory, Joy, author.
Title: Biosphere / Joy Gregory.
Description: New York : Lightbox, [2022] | Series: Earth's systems | Includes index. | Audience: Ages 10-12 | Audience: Grades 4-6
Identifiers: LCCN 2021026199 (print) | LCCN 2021026200 (ebook) | ISBN 9781510553491 (library binding) | ISBN 9781510553507 (ebook)
Subjects: LCSH: Biosphere--Juvenile literature.
Classification: LCC QH343.4 .G74 2022 (print) | LCC QH343.4 (ebook) | DDC 333.95--dc23
LC record available at https://lccn.loc.gov/2021026199
LC ebook record available at https://lccn.loc.gov/20210262003

Printed in Guangzhou, China
1 2 3 4 5 6 7 8 9 0   25 24 23 22 21

072021
111220

Project Coordinator: Priyanka Das
Art Director: Terry Paulhus

Photo Credits
Every reasonable effort has been made to trace ownership and to obtain permission to reprint copyright material. The publisher would be pleased to have any errors or omissions brought to its attention so that they may be corrected in subsequent printings. The publisher acknowledges Alamy, Getty Images, and Shutterstock as its primary image suppliers for this title.